ALGEBRA

Solutions to Common Verbal Problems

Victor P. Vizarra

COPYRIGHT PAGE

ACKNOWLEDGMENTS

I wish to thank all my colleagues, associates and students at Technological Institute of the Philippines who in one way or another inspired me to compile and write this book.

My gratitude also goes to Engr. Serafin Marcaida Jr. and Engr. Mariano Tagayun, both from National Power Corporation, Philippines and to Engr. Stan Villanueva of Batangas who is now residing in Germany, for their support and friendship while we were students at Far Eastern University.

Lastly, my special thanks to my beloved Aida for her support, love and inspiration.

DEDICATION

To all my former and future students of Algebra—

May you cherish this book as much as I do...

PROLOGUE

This is a collection of common verbal problems in Algebra with their corresponding solutions. The topics covered herein cover miscellaneous practical problems that we encounter in our daily lives such as computations of: age, investment, mixture, motion, clock, work related problems leading to linear equations, quadratic equations, simultaneous quadratic equations, variations, arithmetic and geometric progressions, permutations and probability.

Some of the problems were a compilation from the notes and examinations of the author while he was a student at Far Eastern University, Manila and from the lectures he had while teaching Mathematics in the Technological Institute of the Philippines.

The main purpose of this book is to help the students both in high school and college to get a clear understanding of the subject of Algebra by using the problems and solutions presented as a guide in solving similar given problems.

The author appreciates any feedback or suggestions for further improvement of this book.

If you are interested in tutorial services (Frederick, Maryland only), please send e-mail to the author at: engr.vizarra@yahoo.com for reservation.

CONTENTS

I. Verbal Problems Leading to Linear Equations

1. <u>Age Problems:</u>

(1) If Anthony was four times as old as Ben 10 years ago, and if Anthony will be twice as old as Ben 10 years hence, find their ages.

Let x = age of Anthony now,   y = age of Ben now

Ten years ago, their ages were (x - 10) and (y - 10), respectively

$$x - 10 = 4(y - 10)$$
$$x = 4y - 40 + 10$$
$$x = 4y - 30 \quad \text{equation (1)}$$

Ten years hence, their ages will be (x + 10) and (y + 10) respectively

$$x + 10 = 2(y + 10)$$
$$x = 2y + 20 - 10$$
$$x = 2y + 10 \quad \text{equation (2)}$$

Subtracting eq. (2) from eq. (1):

$$x = 4y - 30$$
$$- \ \underline{x = 2y + 10}$$
$$0 = 2y - 40$$
$$y = 20 \text{ years old} \quad \text{(Answer)}$$
$$x = 50 \text{ years old}$$

Check: from equation (1),

$$50 - 10 = 4(20 - 10); \quad 40 = 40$$

from equation (2),

$$50 + 10 = 2(20 + 10); \quad 60 = 60$$

(2) A boy is 1/3 as old as his brother and 8 years younger than his sister. The sum of their ages is 38 years. How old are they?

Let x = age of boy;　$3x$ = age of brother;

$$x + 8 = \text{age of sister}$$

$$x + 3x + (x + 8) = 38$$
$$5x = 30 - 8 = 30$$
$$x = 6 \text{ (age of boy) (Answer)}$$
$$3x = 18 \text{ (age of brother)}$$
$$x+8 = 14 \text{ (age of sister)}$$

2. Investment Problems:

(1) The sum of John's income for one year on $15,000 at simple interest, and on $20,000 for two years at simple interest is $4,800. If the rates of interest were interchanged, he would receive $4,550. Find the rates of interest.

Let $x\%$ and $y\%$ be the respective rates of simple interest.

$$15,000 \cdot x/100 + 2(20,000 \cdot y/100) = 4,800$$
$$150x + 400y = 4,800$$
$$3x + 8y = 96 \text{ equation (1)}$$

15,000 . y/100 + 2 (20,000 . x/100) = 4,550

150y + 400x = 4,550

8x + 3y = 91 equation (2)

Solving equations (1) and (2), we find:

X = 8% and y = 9% (Answer)

(2) The sum of the capitals of Allied, Broker, and Chevy is $110,000. Allied's capital is invested at 7%, Broker's 8%, and Chevy's at 9%, and the sum of their incomes for one year is $9,000. If the rates at which Broker and Chase's investments are interchanged, the income for all is increased by $100. Find their respective capitals.

Let x = Allied's capital

y = Broker's capital

110 -x –y= Chevy's capital

0.07x + 0.08y + 0.09(110,000 – x – y) = 9,000

7x + 8y +990,000 – 9x – 9y = 900,000

2x + y = 90,000   equation (1)

0.07x + .09y + 0.08(110,000 – x – y) = 9100

7x + 9y + 880,000 – 8x – 8y = 910,000

-x + y = 30,000   equation (2)

Subtracting equation (2) from equation (1):

3x = 60,000

3

x = 60,000/3 = $20,000   (Allied's capital)

substituting x = 20,000 in equation (2)

therefore: y = $50,000   (Broker's capital)

110,000 − x − y = $40,000 (Chevy's capital)

3. <u>Mixture Problems</u>

(1) A goldsmith has two alloys of gold, the first being 80% pure gold and the second 70% pure gold. How many ounces of each must be used to make 100 ounces of an alloy which will be 77% pure gold?

Let x = weight in ounces of 80% pure gold

100-x = weight in ounces of 70% pure gold

0.80x = actual gold in 80% pure gold alloy

0.70(100-x) = actual gold in 70% pure gold alloy

0.77(100) = actual gold in mixture

0.80x + 0.70(100-x) = 77

8x + 700 −7x = 770

x = 70 oz. (80% pure gold)

100 − x = 30 oz. (70% pure gold) (Answer)

(2) A chemist has two alcohol solutions of different strengths, 30% alcohol and 45% alcohol solutions, respectively. How many cu.

cm. of each must he use so as to make a make a mixture of 30 cu.

cm. which will contain 36% alcohol?

Let x = volume in cc 0f 30% alcohol

30-x = volume in cc of 45% alcohol

0.30x + 0.45(30-x) = 0.36(30)

30x + 1350 −45x = 1080

-15x = -270

x = 18 c. c. (30% solution)

30 − x = 12 c. c. (45% solution) (Answer)

4. **Motion Problems**

(1) Bill and Jack start at the same time from two places 100 miles apart and travel toward each other. Bill travels 10 miles per hour and Jack 8 miles per hour. If Jack rests 1 hour on the way, in how many hours will they meet?

Let t = time in hours Bill travels

t – 1 = actual time of travel of Jack

10t = distance in miles traveled by Bill

8(t – 1) = distance in miles traveled by Jack

10t + 8(t – 1) = 100

18t = 108

t = 6 hours   (Answer)

(2) A man can row 20 miles downstream in the time it takes him to row 8 miles upstream. He rows downstream for 1-1/2 hours, then turns and rows back for 3 hours, but he finds that he is still 3 miles from his starting place. Find the rate of the man in still water and the rate of the stream.

Solution: Let x = rate of man in still water

y = rate of stream

x + y = rate of man downstream

x – y = rate of man upstream

20/x+y = 8/x-y

$20x - 20y = 8x + 8y$

$\quad 3x = 7y \qquad$ equation (1)

$\dfrac{3}{2}(x+y) = 3(x-y) + 3$

$x + y = 2x - 2y + 2$

$\quad x = 3y - 2$ equation (2)

Solving equation (1) and (2):

$\quad 3(3y-2) = 7y$

$\quad 9y - 6 = 7y$

$\quad 2y = 6; \quad y = 3$ mi./hr.

Substituting y = 3 in equation (2)

$\quad x = 3(3) - 2 = 9 - 2 = 7$ mi./hr. $\quad$ (Answer)

(3) Arthur can walk 4 miles in the time it takes Bradley to walk 5 miles. Arthur requires 3 minutes longer than Bradley to walk a mile. Find their rates.

$\quad$ Let x = rate of Arthur ; y = rate of Bradley

$4/x = 5/y$ or $4/x - 5/y = 0$ equation (1)

$1/x - 1/y = 3/60 = 1/20$

or: $4/x - 4/y = 1/5 \qquad$ equation (2)

Subtracting (2) from (1):

$\quad -1/y = -1/5$

$\quad y = 5$ miles/hr. $\qquad$ (Answer)

$1/x = 1/20 + 1/5 = 5/20; \quad x = 4$ miles/hr.

(4) Ashley runs around a circular track in 60 seconds and Beth in 50 seconds. Five seconds after Ashley starts, Beth starts from the same point in the same direction. When will they be together for the first time, assuming they run around the track continuously?

Let C = circumference of the circular track

t = time when Ashley and Beth will be together for the first time

Reckoned from the time Ashley started.

t – 5 = time of Beth

They will be together when the difference between the distances run is one circumference. Hence,

(t-5) . C/50 – t . C/60 = C

60t – 300 – 50t = 3000

t = 330 seconds = 5 ½ minutes   (Answer)

(5) Arthur and Billy run around a circular track whose circumference is 150 meters. When they run in opposite directions they meet every 5 seconds, but when they run in the same direction from the same point, they are together every 25 seconds. What are their rates?

Let x = rate of Arthur in m/sec.

y = rate of Billy in m/sec.

Equating distances, we have:

5x + 5y = 150;   x + y = 30 equation (1)

25x − 25y = 150;   x − y = 6   equation (2)

x = 18 m/sec.;   y = 12 m/sec.   (Answer)

(6) A and B run two 200-meter races. In the first race A gives B a start of 20 meters and beats him by 5 seconds. In the second race A gives B a start of 8 meters and beats him by 8 seconds. Find their rates.

Let x = rate of A in m/sec.

y = rate of B in m/sec.

$200/x = 180/y − 5$ equation (1)

$200/x = 192/y − 8$ equation (2)

Equating the right members, we have

$180/y − 5 = 192/y − 8$

$−12/y = −3$;   y = 4 m/sec.   (Answer)

Substituting value of y = 4 in equation (1):

$200/x = 180/4 − 5 = 40$

x = 5 m/sec.   (Answer)

(7) A cat is now 50 of her own leaps ahead of a dog, which is pursuing her. How many more leaps will the cat take before it is overtaken if she takes 5 leaps while the dog takes 4, but 2 of the dog's leaps are equivalent to 3 of the cat's leaps?

Let x = additional leaps of the cat

4/5 x = no. of leaps of the dog

$$\frac{x + 50}{4/5 \, x} = 3/2$$

$$x + 50 = 3/2 \cdot 4/5 \, x = 6/5 \, x$$

$$x = 250 \text{ leaps of the cat} \quad \text{(Answer)}$$

(8) A cop is pursuing a thief who is ahead by 72 of his own leaps. The thief takes 6 leaps while the cop is taking 5 leaps, but 4 leaps of the thief are as long as 3 leaps of the cop. How many leaps will each make before the thief is caught?

Let x = additional leaps of the thief

$5/6 \, x$ = no. of leaps of the cop

$$\frac{x + 72}{5/6 \, x} = 4/3$$

$$x + 72 = 4/3 \cdot 5/6 \, x = 10/9 \, x$$

$$x = 648 \text{ leaps of thief}$$

$$5/6 \, x = 540 \text{ leaps of the cop} \quad \text{(Answer)}$$

(9) A battleship started on a 500-mile voyage but was brought to a full stop an hour after starting for military reasons, delaying it a full hour, after which it was ordered to proceed at a reduced velocity equivalent to 75% of its former rate. It arrived at its destination 3 hours and 3 quarters after scheduled time. If it was desired to arrive one hour and a quarter sooner, how far from the destination should it have been stopped?

Let: v = original speed

x = distance from the destination where the ship should have been stopped in order to arrive 1 ¼ hours sooner

From v = d/t, and substituting values based on given situation:

$$2 + \frac{500 - v}{\frac{3}{4}\,v} = 500/v + 3\,\frac{3}{4}$$

$$2 + \frac{2000 - 4\,v}{3\,v} = 500/v + 15/4$$

$$\frac{2\,v + 2000}{3\,v} = \frac{2000 + 15\,v}{4\,v}$$

$$v = 2000/37 = 54\,2/37 \text{ miles/hour}$$

Solving for x:

$$\frac{500 - x}{2000/37} + 1 + \frac{x}{\frac{3}{4}\,(2000/37)} = \frac{500}{2000/37} + 3\,\frac{3}{4} - 1\,\frac{1}{4}$$

$$x = 243\,9/37 \text{ miles} \qquad \text{(Answer)}$$

5. Clock Problems

(1) At what time between 4 and 5 o'clock are the hands of a clock (a) opposite each other, (b) coincident, (c) at right angles?

Let x = no. of minutes after 4 o'clock for each case

$x/12$ = no. of minute marks traveled by hour hand

(a) $x = x/12 + 50$ (b) $x = x/12 + 20$

  $11x = 600$ $11x = 240$

    $x = 54\ 6/11$ min. $x = 21\ 9/11$ min.

  Time = 4:54 6/11 o'clock   Time = 4:21 9/11 o'clock (Answer)

(c) $x = x/12 + 5$ $x = x/12 + 35$

  $11x = 60$ $11x = 420$

    $x = 5\ 5/11$ min. $x = 38\ 2/11$ min.

  Time = 4:5 5/11 o'clock   Time = 4:38 2/11 o'clock (Answer)

(2) It is between 3 and 4 o'clock, and in 20 minutes the minute hand will be as much after the hour hand as it is now behind it. What is the time?

$$15 + x/12 - x + \tfrac{1}{2} \cdot 20/12 = \tfrac{1}{2}(20) - 11/12x =$$

$$10 - 15 - 10/12 = -70/12$$

$$11x = 70$$

$$x = 6\ 4/11 \text{ min.}$$

Time = 3:6 4/11 o'clock   (Answer)

6. Work Problems

(1) Arnold can do a piece of work in p days, while Burt can do the same work in q days. In how many days can they do the work together?

Let x = no. of days both can do the work together

1/p = no. of days Arnold can do the work

1/q = no. of days Burt can do the work

$$1/p + 1/q = 1/x \; ; \quad \underline{p + q} = 1/x$$
$$pq$$

x = pq/(p+q) days   (Answer)

(2) A can do a piece of work in p days, B in q days, and C in r days. In how many days can they do the work together?

Let x = no. of days all can do the work together

$$1/x = 1/p + 1/q + 1/r = \underline{qr + pr + pq}$$
$$pqr$$

x = pqr/(pq + pr + qr) days   (Answer)

(6-3) A tank can be filled by two pipes in 4 and 6 hours, respectively. It can be emptied by a third pipe in 5 hours. In what time can an empty tank be filled if the three pipes are open?

Let x = no. of hours an empty tank can be filled if all three pipes are open.

$1/x = \frac{1}{4} + 1/6 - 1/5 = 13/60$

$\quad x = 60/13 = 4\ 8/13$ hours   (Answer)

II. <u>Simple Quadratic Equations</u>

(1) Solve for x in the quadratic equation: $3x^2 + 2x - 8 = 0$.

(a) By factoring

$$3x^2 + 2x - 8 = 0$$

$$(x + 2)(3x - 4) = 0$$

$$x = -2 \text{ and } 4/3 \qquad \text{(Answer)}$$

(b) By completing the square

$$3x^2 + 2x - 8 = 0$$

$$3x^2 + 2x = 8$$

Divide by 3: $\qquad x^2 + 2/3\, x = 8/3$

Add $(1/3)^2$ to both members: $x^2 + 2/3\, x + (1/3)^2 = 8/3 + 1/9$

Factor the left member: $\qquad\qquad (x + 1/3)^2 = 25/9$

Extract square roots: $\qquad\qquad x + 1/3 = + \text{ or } - (5/3)$

$$x = -1/3 \text{ + or } - 5/3$$

$$x = 4/3 \text{ or } - 6/3$$

Hence, $\qquad\qquad x = 4/3 \text{ and } -2 \quad \text{(Answer)}$

(c) By quadratic formula:

$$3x^2 + 2x - 8 = 0$$

Where: $\qquad a = 3 \quad b = 2 \quad c = -8$

$$x = \frac{-b +/- \sqrt{b^2 - 4\,ac}}{2a}$$

$$= \frac{-2 +/- \text{sq.rt of } (4 - 4(3)(-8)}{2(3)}$$

15

$$= \underline{-2 +/- 10} = 8/6 \text{ or } -12/6$$
$$6$$

$$x = 4/3 \quad \text{and} \quad -2 \qquad \text{(Answer)}$$

(2) Derive the quadratic formula.

Derivation: The formula is derived by completing the square.

$$ax^2 + bx + c = 0$$

Transpose c and divide by a:

$$x^2 + b/a\, x = -c/a$$

To complete the square, we add $(b/2a)^2$ to both members,

$$x^2 + b/a\, x + (b/2a)^2 = -c/a + (b/2a)^2$$

$$(x + b/2a)^2 = -c/a + b^2/4a^2$$

or $\qquad (x + b/2a)^2 = (b^2 - 4ac) / 4a^2$

$$x + b/2a = +/- \underline{(\text{sq. rt. of } b^2 - 4ac)}$$
$$2a$$

$$x = \underline{-b +/- \text{sq. rt. } (b^2 - 4ac)}$$
$$2a \qquad \text{(Answer)}$$

III. Verbal Problems Leading to Quadratic Equations

(1) The sum of the squares of two consecutive positive integers is 265. Find the integers.

Let the consecutive integers be x and (x + 1), then

$$x^2 + (x + 1)^2 = 265$$

$$2x^2 + 2x + 1 = 265$$

$$x^2 + x - 132 = 0$$

$$(x + 12)(x - 11) = 0$$

$$x = -12 \quad \text{and} \quad 11$$

Since the integers are positive, x = 11 and x + 1 = 12   (Answer)

(2) An audience of 540 persons is seated in rows having the same number of persons in each row. If 3 more persons seat in each row, it would require 2 rows less to seat the audience. How many persons were in each row originally?

Let x = no. of persons in each row originally

540/x = no. of rows

$$(x + 3)(540/x - 2) = 540$$

$$540 + 1620/x - 2x - 6 = 540$$

$$1620 - 2x^2 - 6x = 0$$

$$x^2 + 3x - 810 = 0$$

$$(x - 27)(x + 30) = 0$$

$$x = 27 \quad \text{(Answer)}$$

(3) A boat going across a lake 8 miles wide proceeds 2 miles at a certain speed and then completes the trip at a speed ½ mile faster. By doing this, the boat arrives 10 minutes earlier than if the original speed had been maintained. Find the original speed of the boat and the time it took to cross the lake.

Let x = original speed of the boat

t = time to cross the lake

$2/x + 6/(x + ½) = 8/x − 1/6$

$12/(2x + 1) = 6/x − 1/6 = (36 − x)/6x$

$72x = (2x + 1)(36 − x)$

$= -2x^2 + 71x + 36$

$2x^2 + x − 36 = 0$

$(x − 4)(2x + 9) = 0$

x = 4 miles/hour   (Answer)

t = 2/4 + 6/(4 ½) = 11/6 = 1 5/6 hours

Time to cross the lake, t = 1 hour 50 min. (Answer)

(4) Agnes can do a piece of work in 2 days less time than Betty. If both can do the work in 2 2/5 days, how long will it take each to do the work?

Solution: Let x = no. of days Betty can do the work alone

x − 2 = no. of days Agnes can do the work alone

1/x = part of the work Betty can finish in 1 day

1/(x − 2) = part of the work Agnes can finish in 1 day

Hence,   1/x + 1/(x − 2) = 1 /2 2/5 = 5/12

$$12(x - 2) + 12x = 5x(x - 2)$$

$$5x^2 - 34x + 24 = 0$$

$$(x - 6)(5x - 4) = 0$$

$$x = 6, \ 4/5$$

Thus, Agnes can do the work in 4 days and Beth can do the work in 6 days.

IV. Verbal Problems Leading to Simultaneous Quadratic Equations

(1) The perimeter of a rectangle is 68 ft. and its diagonal is 26 ft. Find the area of the rectangle.

Let x & y = the length of the sides

$P = 2x + 2y = 68$

$y = 34 - x$ equation (1)

Sq.rt. of $x^2 + y^2$ = diagonal = 26 ft.

$x^2 + y^2 = 676$ equation (2)

Substituting value of y from eq. (1):

$x^2 + (34 - x)^2 = 676$

$x^2 - 34x + 240 = 0$

$(x - 10)(x - 24) = 0$

$x = 10, \ 24$

If $x = 10, \ y = 24;$ $x = 24, \ y = 10$

In both cases the area = $A = xy = 240$ sq. ft.   (Answer)

(2) Find two numbers such that their sum multiplied by the sum of their squares is 5500, and their difference multiplied by the difference of their squares is 352.

Let x, y be the numbers

$(x + y)(x^2 + y^2) = 5500$ equation (1)

$(x - y)(x^2 - y^2) = 352$ equation (2)

or: $(x + y)(x - y)^2 = 352$ equation (3)

Dividing equation (1) by (3):

$$\frac{x^2 + y^2}{(x - y)^2} = \frac{5500}{352} = 125/8$$

$$8x^2 + 8y^2 = 125x^2 - 250xy + 125y^2$$

$$117x^2 - 250xy + 117y^2 = 0$$

Factoring, $(13x - 9y)(9x - 13y) = 0$

$$X = 9y/13 \text{ or } 13y/9$$

Substituting x = 13y/9 in equation (2):

$$(13y/9 - y)(169/81 \ y^2 - y^2) = 352$$

$$4/9 \cdot 88/81 \ y^3 = 352$$

$$y^3 = 9(81)$$

$$y = 9, \quad x = 13 \quad \text{(Answer)}$$

(3) Two boys, Arnold and Butch are running at constant rates and in the same direction around a circular track whose circumference is 40 ft. Arnold makes one circuit in 2 seconds less time than Butch, and they are together once every minute. Find their rates.

Let x = the rate of Arnold, y = the rate of Butch

$$40/x = 40/y - 2 \quad \text{or} \quad 20x = 20/y - 1 \quad \text{equation (1)}$$

The distances run by the boys in 1 minute should differ by one circumference so that they will be together every minute.

$$60x - 60y = 40$$

$$3x - 3y = 2 \qquad \text{equation (2)}$$

$$y = (3x - 2)/3$$

Substitute value of y in equation (1):

$$20/x = 20/\underline{(3x - 2)} - 1$$
$$3$$

$$= 60/(3x - 2) - 1$$

$$20(3x - 2) = 60x - x(3x - 2)$$

$$3x^2 - 2x - 40 = 0$$

$$(x - 4)(3x + 10) = 0$$

$$x = 4 \text{ ft./sec. } y = 3\ 1/3 \text{ ft./sec.} \quad \text{(Answer)}$$

(4) The difference of the cubes of two positive numbers is 2402 and the cube of their difference is 8. Find the numbers.

Let x, and y be the two positive numbers

$$x^3 - y^3 = 2402 \qquad \text{equation (1)}$$

$$(x - y)^3 = 8 \qquad \text{equation (2)}$$

From equation (2), $x - y = 2$ and $y = x - 2$

Substituting value of $y = x - 2$ in equation (1):

$$x^3 - (x - 2)^3 = 2402$$

$$6x^2 - 12x - 2394 = 0$$

$$x^2 - 2x - 399 = 0$$

$$(x - 21)(x + 19) = 0$$

$$x = 21, \&\ -19$$

Since the numbers are positive, x = 21 and y = 19   (Answer)

V. Variations

(1) The electrical resistance of a cable varies directly as its length and inversely as the square of its diameter. If a cable 2,000 ft. long and ½ inch in diameter has a resistance of 0.1 ohm, find the length of a cable ¾ inch in diameter with a resistance of 1/6 ohm.

Let R = resistance of cable in ohms

L = length of the cable

D = diameter of the cable

$R = k(L/D^2)$; $0.1 = k(2,000/1/4) = 8,000k$

$k = 1/80,000$

Hence, $R = 1/80,000(L/D^2)$ $L = 80,000RD^2$

For R = 1/6 ohm,   we have D = ¾ inch

Therefore: $L = 80,000(1/6)(3/4)^2 = 7,500$ feet   (Answer)

(2) The time required for an elevator to lift a weight varies directly with the weight and the distance through which it is to be lifted and inversely as the power of the motors. If it takes 20 seconds for a 5-HP motor to lift 50 lbs. Through 40 ft. what size of motor is required to lift 1,200 lbs. In 30 seconds through 40 ft.?

Let T = time in seconds

W = weight lifted in lbs.

D = distance in ft.

P = horsepower of motor

T = k(WD/P) ;   20 = k(50(40)/5) = 400k;   k = 1/20

T = 1/20(WD/P) and P = WD/20T

For W = 1,200 lbs., D = 40 ft., and T = 30 seconds, therefore we have

$$P = \frac{1,200(40)}{20(30)} = 80 \text{ HP}$$ (Answer)

(3) For a given mass, the kinetic energy of a rotating wheel varies jointly as the square of its radius and the square of its angular velocity. The kinetic energy of a wheel with a radius of 4 ft. and rotating at the rate of 1 revolution per second is 277 ft-lbs. Find the kinetic energy of a wheel of the same mass whose radius is 5 ft. and which rotates 3 revolutions per second.

Let KE = kinetic energy of the rotating wheel

R = its radius

w = its angular velocity

$KE = kR^2w^2$ $277 = k(4)^2(1)^2 = 16k$;   k = 277/16

For R = 5 ft.,   w = 3 rev./sec., we obtain

$KE = 277/16 \, (5)^2(3)^2 = 3,900$ ft.-lbs.   (Answer)

(4) The horsepower, which a shaft can transmit, varies as the cube of the diameter and the angular speed. If a 3-inch diameter shaft transmits 270 HP when turning 1,000 RPM, find the horsepower which can be transmitted by a 5-inch shaft rotating 1,500 RPM. Find the RPM of a 2-inch shaft if it transmits 100 HP.

Let HP = horsepower of the shaft

D = its diameter   and   w = angular velocity

$HP = kD^3w$ $270 = k(3)^3(1,000) = 27,000k$

$k = 1/100$

$HP = 1/100(D^3)w$ $w = 100HP/D^3$

For D = 5 inches,   w = 1,500RPM   we have:

$HP = 1/100(5)^3(1,500) = 1,875$ HP   (Answer)

For HP = 100   D = 2 inches,   we have:

$w = 100(100)/(2)^3 = 1,250$ RPM   (Answer)

(5) The load of a horizontal beam supported at the ends varies jointly as the breadth and the square of the depth of the beam, and inversely as the length between supports. If a beam 3 inches wide, 8 inches deep, and 16 ft. long can safely support 1,500 lbs., find the length of a beam 6 inches wide, 6 inches deep, and supporting a safe load of 2,250 lbs.

Let P = the load which the beam supports

b = breadth of the beam

d = depth of the beam

L = length of the beam

$P = kbd^2/L;$ $1,500 = k \cdot 3(8)^2/16 = 12k,$ $k = 125$

From $P = kbd^2/L$, therefore $L = 125\, bd^2/P$

For b = 6 inches,   d = 6 inches and   P = 2,250 lbs.,   we have

$L = 125[6(6)^2]/2,250$

$= 12$ feet (Answer)

VI. Arithmetic Progression

(1) Find the sum of the terms of an arithmetic progression having nine terms with the middle term equals to 21.

Let: n = 9   a5= middle term = 21

a5 = a1 +4d = 21   a9 = a1 + 8d

S9 = 9/2[a1 + (a1 + 8d)] = 9/2(2a1 +8d)

= 9(a1 + 4d) = 9(21) = 189(Answer)

(2) What is the sum of the dates in a 31-day month?

Let: a1= 1 a31= 31 n = 31   d = 1

S31 = 31/2(1 + 31) = 31(16) = 496   (Answer)

(3) There are twelve rows of billiard balls in a symmetrical arrangement on a billiard table. If there is one ball in the first row, 3 in the second, 5 in the third, and so on, with two balls more than the preceding one, how many balls are in the 12th row and how many balls in all are there?

Let: The number of balls in A.P.: 1, 3, 5, 7, ...

Therefore, a1 = 1 d = 2 n = 12

A12 = a1 + 11d = 1 + 11(2) = 23   (Answer)

S12 = 12/2 (1 + 23) = 144   (Answer)

(4) To build a dam, 60 men must work 72 days. If all 60 men are employed at the start but the number working is decreased by 5 men at the end of each 12-day period, how long will it take to complete the dam?

Let: Work = 60 men x 72 days = 4,320 man-days

1st 12 days: No. of men = 60 Work done = 60 x 12 = 720

2nd 12 days: No. of men = 55 Work done = 55 x 12 = 660

3rd 12 days: no. of men = 50 Work done = 50 x 12 = 600

The work done form the A.P.: 720, 660, 600,...

$a_1 = 720$ $d = -60$

$a_n = 720 + (n - 1)(-60) = 780 - 60n$

$S_n = 4,320 = n/2 [720 + (780 - 60n)] = 750n - 30n^2$

$30n^2 - 750n + 4,320 = 0$

$n^2 - 25n + 144 = 0$

$(n - 9)(n - 16) = 0$

$n = 9$ or 16

The dam will be completed in 9 x 12 = 108 days   (Answer)

VII. Geometric Progression

(1) If one-fourth of the air in a tank is removed by each stroke of an air pump, find the fractional part of the air remaining after seven strokes of the pump.

Form a table as follows:

Amount of air: 1   ¾ $(3/4)2$ ....

Amount removed: ¼   ¼ . ¾   ¼ . $(3/4)2$ ....

Amount remaining: ¾ $(3/4)2$ $(3/4)^3$ ...

The amount remaining form a G.P. with:

a1 = ¾ r = ¾   n = 7

$a7 = a_1r^6 = ¾ . (3/4)^6 = 3^7/4^7 = 2187/16,384 = 0.1336$ (Answer)

(2) A cask containing 20 gallons of wine was emptied of 1/5 of its contents and then filled with water. If this is done 6 times, how many gallons of wine remain in the cask?

Solution: 1st drawing: portion of wine remaining = $1 - 1/5 = 4/5$

2nd drawing: portion remaining = $4/5(\underline{1 - 1/5}) = (4/5)^2$

1

3rd drawing: portion remaining = $(4/5)^2(\underline{1 - 1/5}) = (4/5)^3$

1

These form a G. P. with $a_1 = 4/5$ r = 4/5

For n = 6, portion remaining = $a_1r^5 = (4/5)(4/5)^5 = 0.2621$

Qty. of wine remaining in the cask = 0.2621 x 20 gals.

$$= 5.242 \text{ gallons} \quad \text{(Answer)}$$

(3) A pendulum is brought to rest by air resistance, each swing being 11/12 as much as the preceding one. If the lower end of the pendulum describes an arc 24 inches long in the first swing, what will be the total length of the path which the pendulum describes before it comes to rest?

The lengths of the arcs described by the pendulum in successive swings will be:

$$24, 11/12(24), \quad (11/12)^2(24), \quad (11/12)^3(24) \dots$$

These form an infinite G. P. with $a_1 = 24$, $\quad r = 11/12$

Total length of the arc described = $24/(1 - 11/12) = 288$ inches

(Answer)

(4) An equilateral triangle is inscribed within a circle whose diameter is 12 inches. In this triangle, a circle is inscribed; and in this circle, another equilateral triangle is inscribed, and so on indefinitely. Find (a) the sum of the perimeters of all the triangles, and (b) the sum of the areas of all the triangles.

Solution:

(a) Perimeters: (b) Areas

   P1 = perimeter of ABC   A1 = area of ABC

     = $3(6\sqrt{3}) = 18\sqrt{3}$ inches   = $\frac{1}{2}(6\sqrt{3})9 = 27\sqrt{3}$

P2 = perimeter of DEF   A2 = area of DEF

= $9\sqrt{3}$ inches   = $\frac{1}{2}(3\sqrt{3})(9/2) = (27\sqrt{3})/4$

r = P2/P1 = ½   r = A2/A1 = 1/4   P = total

perimeter   A = total area

= $(18\sqrt{3})/(1 - \frac{1}{2})$ = $(27\sqrt{3})/(1 - \frac{1}{4})$

= $36\sqrt{3}$ inches   = $36\sqrt{3}$ sq. inches

(Answer)

= 62.35 inches   = 62.35 sq. inches

VIII. <u>Permutations</u>

(1) How many 3-digit numbers can be formed from the digits 2, 4, 6, 8 and 9 if (a) no digit is repeated in any number? (b) if repetitions are allowed?

Solution: (a) The first digit can be filled in 5 ways, the second in 4 ways, and the third in 3 ways. Therefore, the 3-digit number can be formed in:

$\quad$ 5 . 4 . 3 = 60 ways $\quad$ (Answer)

(b) Since repetitions are allowed, each of the three digits can be filled in 5 ways. Therefore, the 3-digit number can be formed in:

$\quad$ 5 . 5 . 5 = 125 ways $\quad$ (Answer)

(2) A building has 6 outside doors. In how many ways can a person enter and leave (a) by any door? (b) by a different door?

Let: (a) The person can enter or leave 6 doors, each in 6 ways. Therefore, he can enter and leave in: 6 . 6 = 36 ways (Answer)

(b) The person can enter through any of the 6 doors, but can leave only through 5 doors. Therefore: 6 . 5 = 30 ways (Answer)

(3) In how many ways can 6 persons be seated in a room where there are 9 seats?

Let A = 1st person : can occupy any of 9 seats

B = 2nd person: can occupy any of 8 seats

C = 3rd person: can occupy any of 7 seats

D = 4th person: can occupy any of 6 seats

E = 5th person: can occupy any of 5 seats

F = 6th person: can occupy any of 4 seats

Therefore, they can be seated in: 9 . 8 . 7 . 6 . 5 . 4 = 60,480 ways

(Answer)

(4) An engineering club has a membership of 24 Civil Engineers, 20 Chemical Engineers, 16 Mechanical Engineers and 12 Electrical Engineers. If a committee of 4 members, one from each group, is to be formed, how many such committees can be formed?

Solution: The committee of 4 members from each group can be formed in:

24 . 20 . 16 . 12 = 92,160 ways   (Answer)

(5) Evaluate: (a) P(8, 1)   (b) P(8, 4)

Solution: P(n, r) = n!/(n-r)!   P(n, n) = n!

(a) P(8, 1) = 8!/7! = 8   (Answer)

(b) P(8, 4) = 8!/4! = 5 . 6 . 7 . 8 = 1,680 (Answer)

(6) Find the number of permutations which can be formed from all letters of:

(a) CALIFORNIA (b) MARYLAND (c) PHILIPPINES (d) MARINDUQUE

Solution: (a) CALIFORNIA: 10 letters,   2 A's   2 I's

   $P = 10!/2!\ 2! = 3,628,800/4 = 907,200$ (Answer)

   (b) MARYLAND: 8 letters, 2 A's

   $P = 8!/2! = 40,320/2 = 20,160$ (Answer)

   (c) PHILIPPINES: 11 letters,   3 P's   3 I's

   $P = 11!/3!\ 3! = 39,916,800/36 = 1,108,800$ (Answer)

   (d) MARINDUQUE: 10 letters,   2 U's

   $P = 10!/2! = 3,628,800/2 = 1,814,400$ (Answer)

(7) There are 2 copies of physics books, 3 copies of math book, and 4 copies of chemistry book with covers of different colors for each kind of book. In how many different ways can they be placed on a shelf?

Solution: There are 9 books total, 2 physics book, 3 math book, 4 chemistry book

Hence, they can be placed on a shelf in:

   $P = 9!/2!\ 3!\ 4! = 362,880/(2 \times 6 \times 24) = 362,880/288 = 1,260$ ways (Ans.)

(8) Find the number of ways 2 dollars, 4 quarters and 6 dimes can be given to 12 children, if each child gets a coin?

Solution: P = 12!/(2! 4! 6!) = 7 . 8 . 9 . 10 . 11 . 12/(2 x 24) = 13,860 ways (Ans.)

(9) A man and his wife invited 6 of their friends to dinner. They are to be seated around a round table. How many seating arrangements can be made (a) if the man and his wife are to sit side by side? (b) if the man and his wife sit opposite each other? (c) if the 8 people can sit in any way they please?

Solution: (a) The man and his wife can sit in 2! Ways, while the 6 guests can sit in 6! ways.
Hence the 8 people can be seated in 2! 6! = 2(720) = 1,440 ways
   (Answer)
   (b) They can also be seated in 2! 6! = 2(720) = 1,440 ways
   (Answer)
   (c) The 8 people can be seated in any way they please in 7! ways
      7! = 5,040 ways (Answer)

IX. **Probability**

(1) From a box containing 4 white balls, 6 blue balls and 10 black balls, one ball is drawn at random. Determine the probability that it is (a) white, (b) not white, (c) blue, (d) blue or white.

Let: The total number of balls = 4 + 6 + 10 = 20

(a) p = ways of drawing 1 out of 4 white balls/ways of drawing 1 out of 20 balls

   = 4/20 = 1/5   (Answer)

(b) p = 1 − 1/5= 4/5   (Answer)

(c) p = 6/20 = 3/10   (Answer)

(d) p = 1/5 + 3/10 = 5/10 = ½ (Answer)

(2) Dice have their faces numbered from 1 to 6. Determine the probability of throwing a total of 7 in a single throw with two dice.

Let: The total number of ways in which the six faces of one die can be associated with any of the 6 faces of the other die = 6 . 6 = 36

A 7 can be thrown in 6 ways: 1, 6; 2, 5; 3, 4; 4, 3; 5 , 2; 6, 1...

Therefore: p = 6/36 = 1/6   (Answer)

(3) A pack of cards contains 52 cards: 13 spades, 13, clubs, 13 hearts, and 13 diamonds. A suit consists of all the 13 cards of the same kind. Of the 52 cards, 4 are aces, one from each suit. The hearts and diamonds are colored red, the spades and clubs black. Four cards are drawn from a pack, each card being returned to the pack before the next card is drawn. Determine the probability that all are (a) spades, (b) aces, (c) red cards.

Solution: The probability that a spade is drawn from the pack each time is:

13/52 = ¼. Hence, if a card is drawn 4 times, the probability that all will be spades is: $p = (1/4)^4 = 1/256$ (Answer)

(b) Since there are 4 aces:

$p = (4/52)^4 = (1/13)^4 = 1/28,561$ (Answer)

(c) Since there are 26 red cards:

$p = (26/52)^4 = (1/2)^4 = 1/16$ (Answer)

(4) A man bought 10 Lotto tickets with a jackpot prize of $2,000,000. If there are a total of 4,000 tickets, what is his mathematical expectation?

Probability of winning = 10/4,000 = 1/400

Expectation = probability of winning x sum of money

= 1/400($2,000,000) = $5,000 (Answer)

(5) A coin is tossed 5 times. Determine (a) the probability of getting at least 3 heads (b) the odds in favor of getting at least 3 heads

Each time the coin is tossed:

Probability of a head = probability of a tail = ½

Therefore: (a) Probability that 3 of the 5 tosses will be heads equals:

$P = (1/2)^3 = 1/8$

Probability that the other 2 tosses will be tails:

$P = (1/2)^2 = ¼$

Since C(5, 3) different selections of 3 can be made from tosses, the probability that exactly 3 will be heads, is:

$C(5, 3)(1/2)^3(1/2)^2 = 10(1/8)(1/4) = 5/16$

Probability of exactly 4 heads = $C(5, 4)(1/2)^5 = 5/32$

Probability of exactly 5 heads = $(1/2)^5 = 1/32$

Hence, P = 1/32 + 5/32 + 5/16 = 16/32 = ½   (Answer)

(b) The odds in getting at least 3 heads is 2:1 or 2.   (Answer)

X. Miscellaneous Practical Problems:

(1) A marksman fires at a target 420 meters away and hears the bullet strike 2 seconds after he pulled the trigger. An observer 525 meters away from the target and 455 meters from the marksman hears the bullet strike the target one second after he hears the report of the rifle. Find the velocity of the bullet and the velocity of sound.

Let V_b = velocity of the bullet

V_s = velocity of the sound

420/Vb + 420/Vs = 2   equation (1)

455/Vs = 420/Vb + 525/s – 1 equation (2)

From eq. (1) and eq. (2):

420/Vb = 2 – 420/Vs = 1 – 70/s

350/Vs = 1;   Vs = 350 meters/sec. (Answer)

420/Vb = 1 – 70/Vs = 1 – 1/5 = 4/5

Vb = 525 meters/sec   (Answer)

(2) Capital One has invested $5,000,000 in three different transactions. First, in real estate earning 9% interest per annum; second is in loans earning 6% per annum and the third investment is in municipal bonds earning 4% per annum. The total annual income in interests is $340,000 but the annual interest in loans is 3 times that in bonds. How much is each investment?

Let x = amount invested in real estate

y = amount invested in loans

z = 5,000,000 − x − y = amount invested in bonds

0.09x + 0.06y + 0.04(5,000,000 − x − y) = 340,000

9x + 6y + 20,000,000 − 4x − 4y = 34,000,000

5x + 2y = 14,000,000   equation (1)

0.06y = 3(0.04)(5,000,000 − x − y)

6y = 60,000,000 −12x − 12y

12x + 18y = 60,000,000

2x + 3y = 10,000,000   equation (2)

Solving equations (1) and (2) simultaneously,

x = $2,000,000 y = $2,000,000

z = $5,000,000 − x − y = $1,000,000 (Answer)

(3) In Deep Underground Poetry, there's a mix of Australians, Americans, Greeks, Germans and British. The Australians are one less than 1/3 of the British, and three less than half the Americans; the British and the Germans outnumber the Greeks and Americans by 3; the Greeks and British form one less than half the group, and the Greeks and Americans form 7/16 of the group. How many persons of each nationality were there?

Denote the number of Australians, Germans, Greeks, British and Americans by x, y, z, u, v respectively; then we have

x = 1/3u − 1 or 3x = u − 3 equation (1)

x = 1/2v – 3   or   2x = v – 6 equation (2)

y + u – z – v = 3   equation (3)

z + u = ½(x + y + z + u + v) – 1   or   x + y – z – u + v = 2   equation (4)

z + v = 7/16(x + y + z + u + v)   or   7x + 7y – 9z + 7u – 9v = 0   equation (5)

Subtracting eq. (4) from eq. (3), we get: 2u – 2v – x = 1

From eq. (1) and eq. (2):   u = 3x + 3 and   v = 2x + 6

Hence,   6x + 6 – 4x – 12 – x = 1

$$x = 7$$

u = 3(7) + 3 = 24;   v = 2(7) + 6 = 20

From eq. (3) and eq. (5):   y + 24 – z – 20 = 3

y – z = 1   and   9z – 7y = 37

Thus   y = 14 and   z = 15

Therefore: Australians = 7   Germans = 14   Greeks = 15

British = 24   Americans = 20 (Answer)

(4) During the last 2013 election, the total number of votes recorded in a certain municipality was 8,600. Had one-third of Lord Allan Velasco's supporters stayed away from the polls and one-half of Regina Reyes' behaved likewise, Velasco's majority votes would have been reduced by 200. How many votes did Velasco and Reyes actually receive?

Let x = the number of votes for Velasco

8,600 – x = the number of votes for Reyes

x – (8,600 – x) = 2x – 8600 = Velasco's majority over Reyes

$2/3\,x - \tfrac{1}{2}\,(8{,}600 - x) = 2x - 8{,}600 - 200$

$\qquad 7/6\,x - 4{,}300 = 2x - 8{,}800$

$\qquad\qquad 5/6\,x = 4{,}500$

$\qquad\qquad\quad x = 5{,}400$ votes [for Velasco] (Answer)

$\qquad\quad 8{,}600 - x = 3{,}200$ votes [for Reyes] (Answer)

(5) B&G Co. and DBA Construction submitted separate proposals for the construction of stainless steel DC dryers and blanching equipment, with B&G offering the lower price for the winning bid. Had B&G and DBA reduced their bid prices by 5% and 10% respectively, B&G still would have won the bid but the difference in their bids would have been reduced by $3,000.00. If the sum total of the bids is $90,000.00, what are the bids of each contractor?

Let x = amount of B&G's bid

$90{,}000 - x$ = amount of DBA's bid

$(90{,}000 - x) - x = 90{,}000 - 2x$ = difference in the bids

$0.95x$ = reduced bid of B&G (5%)

$0.90(90{,}000 - x)$ = reduced bid of DBA (10%) = $81{,}000 - 0.90x$

New difference in bids = $(81{,}000 - 0.90x) - 0.95x = 81{,}000 - 1.85x$

$\qquad 81{,}000 - 1.85x = (90{,}000 - 2x) - 3{,}000$

$\qquad\qquad\qquad = 87{,}000 - 2x$

$\qquad\qquad 0.15x = 6{,}000$

$\qquad\qquad\quad x = \$40{,}000.00$ [B&G's bid] (Answer)

$\qquad\quad 90{,}000 - x = \$50{,}000.00$ [DBA's bid] (Answer)

(6) B&G has 50 men of the same capacity at work on a job. They can complete the job in 30 days, 8 hours per day, but the contract expires in 20 days. B&G decides to put 20 additional men. If all the men get $10.00 per day and the liquidated damages are $500.00 per day over his contract time, will he complete the job on time or if not, would B&G save money by putting enough men to complete the job on the required time?

Work = 50 men x 30 days x 8 hours/day = 12,000 man-hours
If B&G adds 20 more men, work can be completed by 70 men in:

12,000/70 x 8 = 21 3/7 or 22 days

Therefore the contractor can't complete the job on time by putting 20 more men.

That would be 2 days over the contract completion time equivalent to $1000.00 liquidated damages (@$500.00/day). To finish the job on time he should put 25 additional men.

In 30 days:	In 20 days:
Wages = 50 x 30 x 10 = $15,000.00	75 x 20 x 10 = $15,000.00
Damages = 500 x 10 = $5,000.00	None
Total cost = $20,000.00	Total cost = $15,000.00

Hence B&G saves $5,000.00 by putting more men on the job to finish the work on time.

(7) Mr. Rosales has 408 animals in his farm consisting of turkeys, pigs and chickens. His turkeys are 4 times as many as the pigs

lacking 8 and his chickens are 5 times as many as the pigs lacking 4. How many animals of each kind are there?

Let x = number of turkeys

y = number of pigs

z = number of chickens

Total number of animals = $x + y + z = 408$ equation (1)

$x = 4y - 8$ equation (2)

$z = 5y - 4$ equation (3)

Substituting the values of x and z in equation (1) we have

$(4y - 8) + y + (5y - 4) = 408$

$10y = 420$

$y = 42$ (number of pigs)

Substituting value of y = 42 in equations (2) and (3) we have

$x = 4(42) - 8 = 160$ (number of turkeys)

$z = 5(42) - 4 = 206$ (number of chickens)

(8) If Mr. Smith's rate of doing a piece of work as compared to Mr. Anderson is 2:3, how long will it take Mr. Anderson to do it if Mr. Smith does a piece of work in 30 days?

Since Mr. Smith can do the work in 30 days, his rate is 1/30;

If Mr. Anderson can do the work in x days, then his rate is 1/x.

Thus, $1/30 : 1/x = 2 : 3$

$2/x = 3/30$

$x = 60/3 = 20$ days (Answer)

(9) In what time would Allan, Ben and Carl do a work together if Allan could do it in 6 hours more, Ben alone in 1 hour more, and Carl alone in twice the time?

Let x = number of hours working together

x + 6 = number of hours Allan can do the work alone

x + 1 = number of hours Ben can do the work alone

2x = number of hours Carl can do the work alone

Therefore working together they can do

1/x + 6 + 1/x +1 + 1/2x in one hour but they can also do 1/x of the work in 1 hour. Hence,

$$1/(x+6) + 1/(x+1) + 1/2x = 1/x$$

Simplifying, $3x^2 + 7x - 6 = 0$

Solving, $(3x - 2)(x + 3) = 0$

x = 2/3 hour or 40 mins.   (Answer)

(10) Find the numbers such that their sum multiplied by the sum of their squares is 65, and their difference multiplied by the difference of their squares is 5.

Let x and y be the numbers

$(x + y)(x^2 + y^2) = 65$ equation (1)

$(x - y)(x^2 - y^2) = 5$ equation (2)

Dividing eq. (1) by eq. (2), we get

$$\frac{x^2 + y^2}{(x - y)^2} = 13$$

$x^2 + y^2 = 13x^2 - 26xy + 13y^2$

$6x^2 - 13xy + 6y^2 = 0$

$(2x - 3y)(3x - 2y) = 0$

x = 3/2y or x = 2/3y

Substituting values of x in equation (2) we get,

If x = 3/2y, then $(y/2)(5/4y^2) = 5$

$y^3 = 8$ y = 2 and   x = 3

If x = 2/3y, then $(-y/3)(-5/9y^2) = 5$

$Y^3 = 27$ y = 3 and   x = 2

Hence the numbers are 3 and 2.   (Answer)

(11) The sides of a right triangle are in arithmetic progression whose common difference is 6. Find the value of the sides.

Let the sides be: $c = a + 6$ (hypotenuse)

$\qquad a = a$ (base)

$\qquad b = a - 6$ (height)

Using Pythagorean theorem:

$c^2 = a^2 + b^2$

$(a + 6)^2 = a^2 + (a - 6)^2$

Simplifying we get,

$a^2 - 24a = 0;\qquad a = 0$ or 24

Hence the sides are: $a = 24$ $b = 18$ $c = 30$ (Answer)

(12) Two members of the high school track and field team, Jack and Paul run at constant speeds along a circular track 1,350 meters in circumference. Running in opposite directions, they meet every 3 minutes. While running in the same direction, they are together every 27 minutes. Find their speeds in feet per second.

Let x = time in minutes for Jack to run one circumference

  y = time in minutes for Paul to run one circumference

$1/x + 1/y = 1/3$ equation (1)

$1/x - 1/y = 1/27$ equation (2)

$2/x = 1/3 + 1/27 = 10/27$;   x = 5.4 minutes

$2/y = 1/3 - 1/27 = 8/27$;   y = 6.75 minutes

Speed = distance/time

  Speed of Jack = 1,350/5.4 = 250 meters/min. x 3.28/60

  = 13.67 feet/second   (Answer)

  Speed of Paul = 1,350/6.75 = 200 meters/min. x 3.28/60

  = 10.93 feet/second (Answer)

ABOUT THE AUTHOR

Victor P. Vizarra graduated from Far Eastern University, Manila with a degree of Bachelor of Science in Electrical Engineering. After graduation he taught College Algebra, Geometry, Trigonometry, Differential Equations, Physics and Electrical Engineering subjects in Technological Institute of the Philippines in Manila for two years.

He had six years of project engineering and equipment maintenance experience prior to migrating to the United States in the late eighties. He is a Technology Consultant, poet/blogger and an advocate for solar, wind and other renewable forms of energy.

www.ingramcontent.com/pod-product-compliance
Lightning Source LLC
Chambersburg PA
CBHW021443170526
45164CB00001B/365